Exploring Biomes

GRASSLAND BIOMES

by Lela Nargi

Ideas for Parents and Teachers

Pogo Books let children practice reading informational text while introducing them to nonfiction features such as headings, labels, sidebars, maps, and diagrams, as well as a table of contents, glossary, and index.

Carefully leveled text with a strong photo match offers early fluent readers the support they need to succeed.

Before Reading

- "Walk" through the book and point out the various nonfiction features. Ask the student what purpose each feature serves.
- Look at the glossary together. Read and discuss the words.

Read the Book

- Have the child read the book independently.
- Invite him or her to list questions that arise from reading.

After Reading

- Discuss the child's questions. Talk about how he or she might find answers to those questions.
- Prompt the child to think more. Ask: Many hoofed mammals live in grasslands. Why do you think this is?

Pogo Books are published by Jump!
5357 Penn Avenue South
Minneapolis, MN 55419
www.jumplibrary.com

Copyright © 2023 Jump!
International copyright reserved in all countries. No part of this book may be reproduced in any form without written permission from the publisher.

Library of Congress Cataloging-in-Publication Data

Names: Nargi, Lela, author.
Title: Grassland biomes / by Lela Nargi.
Description: Minneapolis, MN: Jump!, Inc., [2023]
Series: Exploring biomes | Includes index.
Audience: Ages 7-10
Identifiers: LCCN 2021059781 (print)
LCCN 2021059782 (ebook)
ISBN 9781636907598 (hardcover)
ISBN 9781636907604 (paperback)
ISBN 9781636907611 (ebook)
Subjects: LCSH: Grassland ecology–Juvenile literature.
Classification: LCC QH541.5.P7 N37 2023 (print)
LCC QH541.5.P7 (ebook)
DDC 577.4–dc23/eng/20211209
LC record available at
https://lccn.loc.gov/2021059781
LC ebook record available at
https://lccn.loc.gov/2021059782

Editor: Eliza Leahy
Designer: Emma Bersie

Photo Credits: kavram/Shutterstock, cover (left); Fallsview/Dreamstime, cover (right); Konstantin Shishkin/Shutterstock, 1; irakite/Shutterstock, 3; Lawrence Cruciana/Shutterstock, 4; Vadim Petrakov/Shutterstock, 5; Guaxinim/Shutterstock, 6-7; Znm/Dreamstime, 8-9; Tatyana Andreyeva/Shutterstock, 10; Danita Delimont/Alamy, 11; A_Lesik/Shutterstock, 12-13t; R.Moore/Shutterstock, 12-13b; MollieGPhoto/Shutterstock, 14-15tl; aleksander hunta/Shutterstock, 14-15tr; Chantal de Bruijne/Shutterstock, 14-15bl; slowmotiongli/Shutterstock, 14-15br; PHOTOCREO Michal Bednarek/Shutterstock, 16-17tl; Knumina Studios/Shutterstock, 16-17tr; Foto 4440/Shutterstock, 16-17bl; Cephas Picture Library/Alamy, 16-17br; Krasowit/Shutterstock, 18; ZhakYaroslav/Shutterstock, 19; Lorcel/Shutterstock, 20-21; Independent birds/Shutterstock, 23.

Printed in the United States of America at Corporate Graphics in North Mankato, Minnesota.

TABLE OF CONTENTS

CHAPTER 1
Wide Open Spaces............................4

CHAPTER 2
Life in the Grass.............................10

CHAPTER 3
Grasslands and Us..........................18

ACTIVITIES & TOOLS
Try This!......................................22
Glossary......................................23
Index..24
To Learn More...............................24

CHAPTER 1
WIDE OPEN SPACES

How did the grassland **biome** get its name? Many kinds of grasses grow in grasslands. They also have many kinds of wildflowers.

wildflowers

Grasslands are wide open spaces. Most have few trees. They get 20 to 35 inches (51 to 89 centimeters) of **precipitation** each year. That is more than a desert but less than a forest.

CHAPTER 1   5

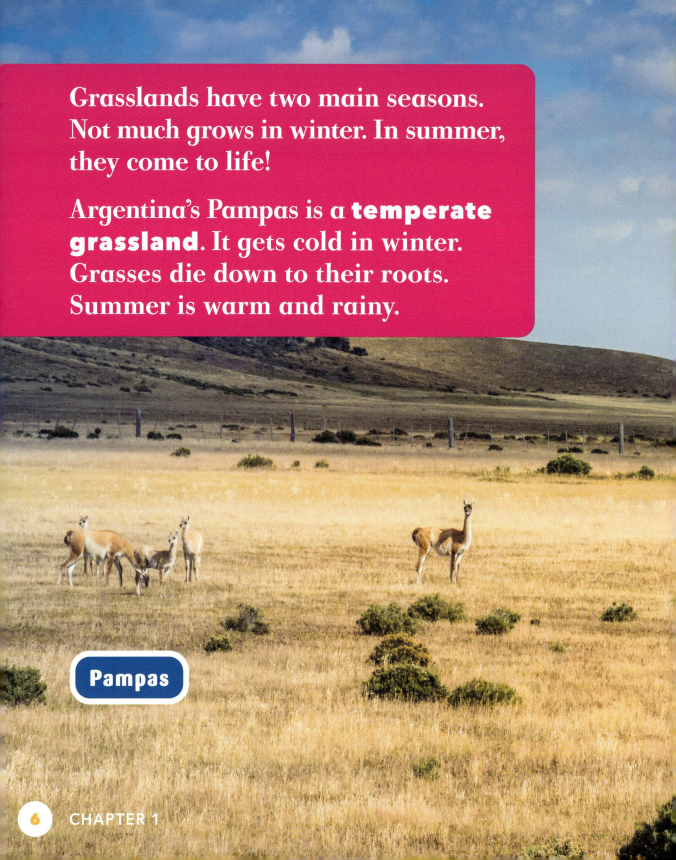

Grasslands have two main seasons. Not much grows in winter. In summer, they come to life!

Argentina's Pampas is a **temperate grassland**. It gets cold in winter. Grasses die down to their roots. Summer is warm and rainy.

Pampas

TAKE A LOOK!

Grasslands cover about 25 percent of Earth's land. Take a look!

= grassland

CHAPTER 1

The Serengeti in Tanzania is a savanna. Savannas are **tropical grasslands**. They also have two main seasons. Summer is warm and wet. Winter is cool but not cold.

DID YOU KNOW?

Grasslands have many names. These include prairie, steppe, savanna, veld, and pampas.

CHAPTER 2
LIFE IN THE GRASS

Many **species** of grasses grow in grasslands. Grasses on the dry Russian steppe are short.

Russian steppe

Grasses are tall in the wet Joseph H. Williams Tallgrass Prairie **Preserve** in Oklahoma. Flowers **bloom** as far as you can see. Insects have plenty to eat.

CHAPTER 2

Wildfires sometimes spread across grasslands. These keep the land healthy. How? They burn dead leaves and grasses. This lets in more sunlight. It makes room for new plants to grow.

DID YOU KNOW?

Grasses in this biome have long roots. This helps them when it is dry. Long roots reach water deep in the ground.

CHAPTER 2 13

elk

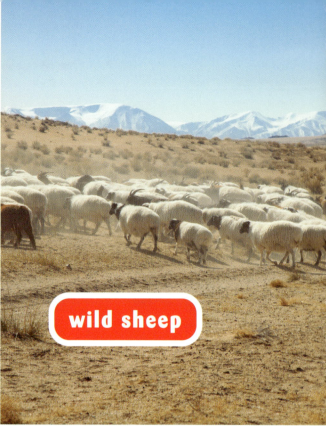
wild sheep

zebras

tapirs

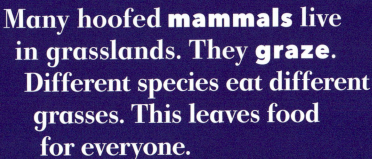

Many hoofed **mammals** live in grasslands. They **graze**. Different species eat different grasses. This leaves food for everyone.

Elk and mule deer trot on the Great Plains. Antelope and wild sheep live on the steppes. Wildebeest and zebras roam East Africa's savanna. In the Pampas, there are tapirs and deer.

CHAPTER 2   15

Lions hunt giraffes in South Africa's veld. Prairie dogs dig **burrows** on the Canadian Prairie. Foxes and armadillos prowl the Pampas. Kangaroos hop on the Australian savanna.

CHAPTER 2

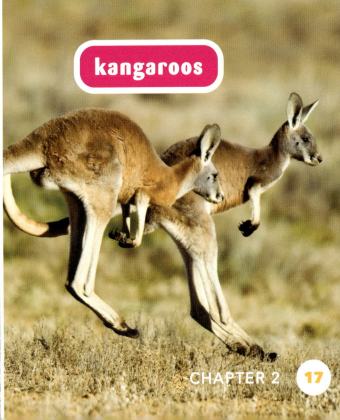

CHAPTER 2 17

CHAPTER 3
GRASSLANDS AND US

Grasslands help slow **climate change**. How? **Carbon dioxide** heats Earth. Grasses take it from the air. They store it in their roots and in the soil. This helps keep Earth cool.

roots

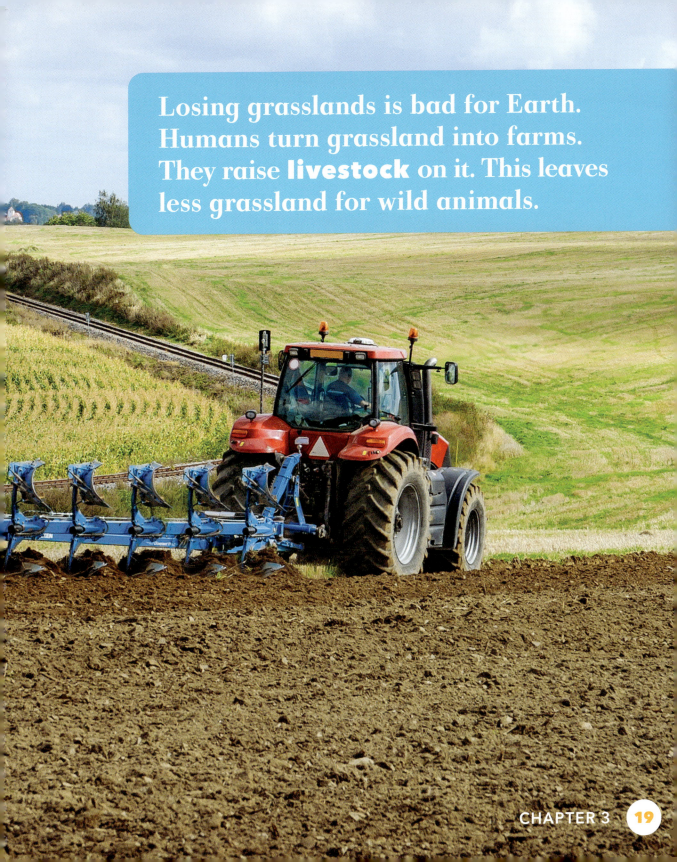

Losing grasslands is bad for Earth. Humans turn grassland into farms. They raise **livestock** on it. This leaves less grassland for wild animals.

CHAPTER 3 · 19

National parks protect many grasslands. We visit them. We learn about these important biomes. In the western United States, Yellowstone National Park has 200 animal species. Many, including bison, elk, and sandhill cranes, live in the park's grasslands.

Would you like to visit a grassland? What animals would you like to see?

DID YOU KNOW?

Millions of bison once roamed North America. By the late 1800s, they were hunted nearly to **extinction**. Today, their numbers are growing.

CHAPTER 3

ACTIVITIES & TOOLS

TRY THIS!

PLANT MINI GRASSLANDS

How does precipitation affect how grasslands grow? Find out in this interesting activity!

What You Need:

- permanent marker
- two small aluminum trays with a few small holes poked in the bottom of each
- soil
- packet of meadow grass seeds or wildflower seeds
- watering can or spray bottle
- water
- plastic wrap

1. Using a permanent marker, write #1 on the first tray. This will be your temperate grassland. Write #2 on the second tray. This will be your tropical grassland.

2. Spread a thick layer of soil over the bottom of each tray.

3. Following the seed packet's instructions, plant the grass or wildflower seeds. Water them according to the instructions. Check how much sunlight they need, and put them in a spot that follows these instructions.

4. Cover the trays with plastic wrap. Keep the soil damp until the seeds sprout.

5. Once they sprout, water tray #1 every time the soil is dry. Water tray #2 every day for a week. Then stop watering tray #2.

6. After a few weeks, which tray has taller, fuller grasses? Why do you think this is?

GLOSSARY

biome: A habitat and everything that lives in it.

bloom: To produce flowers.

burrows: Tunnels or holes in the ground that are made or used as homes by animals such as prairie dogs and rabbits.

carbon dioxide: A gas that is a mixture of carbon and oxygen, with no color or odor.

climate change: Changes in Earth's weather and climate over time.

extinction: The fact or process of a group of living things ceasing to exist.

graze: To feed on grass.

livestock: Animals that are kept or raised on a farm or ranch.

mammals: Warm-blooded animals that have hair or fur and usually give birth to live babies.

precipitation: The falling of water from the sky in the form of rain, sleet, hail, or snow.

preserve: A place where plants and animals are protected in their natural environment.

species: One of the groups into which similar animals and plants are divided.

temperate grassland: A large, open, flat, grassy area with cold winters and warm summers.

tropical grasslands: Large, open, flat, grassy areas with dry and wet seasons and mild temperatures year-round.

wildfires: Fires in wild areas that are not controlled by humans.

ACTIVITIES & TOOLS

INDEX

animals 15, 16, 19, 21
Canadian Prairie 16
farms 19
grasses 4, 6, 10, 11, 12, 15, 18
Great Plains 15
insects 11
Joseph H. Williams Tallgrass Prairie Preserve 11
Pampas 6, 15, 16
prairie 9, 11, 16
precipitation 5
roots 6, 12, 18
savanna 9, 15, 16
Serengeti 9
soil 18
steppe 9, 10, 15
summer 6, 9
temperate grassland 6
trees 5
tropical grasslands 9
veld 9, 16
wildfires 12
wildflowers 4, 11
winter 6, 9
Yellowstone National Park 21

TO LEARN MORE

Finding more information is as easy as 1, 2, 3.

1. Go to www.factsurfer.com
2. Enter "grasslandbiomes" into the search box.
3. Choose your book to see a list of websites.

24 ACTIVITIES & TOOLS